Quadcopter Physical Model, Axis and GPS Control

Impressum:

Bibliographische Information der Deutschen Nationalbibliothek:
Die Deutsche Nationalbibliothek verzeichnet diese Publikation in der Deutschen Nationalbibliografie; detaillierte bibliografische Daten sind im Internet über http://dnb.dnb.de abrufbar.

© 2021 Roland Büchi

Herstellung und Verlag: BoD – Books on Demand, Norderstedt
ISBN: 978-3-7534-9104-2

Contents

Preamble

This booklet is intended for everyone who wants to set the control parameters of a quadcopter or who works on his own projects with a quadcopter. For anyone who wants to understand why a quadrocopter flies at all, or how it realizes the self-balancing, and how to choose controller parameters so that it reaches a predetermined angle of attack in nick or roll, the reading is recommended.

For this purpose, the flight mechanics is first derived in the chapter 'Physical Model'. This leads to the simple model of an axis with two motors. So that a quadcopter can move in one direction, a regulation must be provided which brings it into an angle of attack φ. This is dealt with in the 'Axis Control' chapter. The drift itself is then either controlled visually by the quadcopter pilot himself or with a superimposed controller that uses GPS as a sensor. This regulation is dealt with in the chapter 'GPS Control'.

There are many controller structures for quadcopters in the literature. The one discussed here was deliberately chosen so that it can be implemented with as many flight controllers as possible. In particular, the separation into a subordinate axis control and a superimposed GPS control corresponds to the state of the art in quadcopter controls.

1   Physical Model

In this chapter a model of the axes is developed. For this purpose, the forces acting are first derived using the flight mechanics. A physical model of the system is then calculated.

1.1   Hover flight

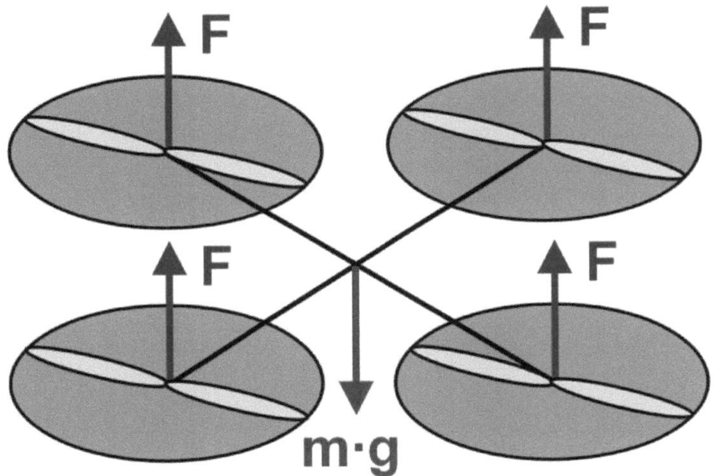

Figure 1: Forces in hover flight

Figure 1 shows a force diagram for normal hovering. From physics 'actio and reactio' is known, and the same can be applied here. In hovering flight namely the sum of the four upward forces is equal to the downward force. The upward-pointing forces F are created by the four propellers and called 'thrust forces' and the downward-pointing force is the weight force (mass m multiplied by the acceleration of gravity g). So:

$$4 \cdot F = m \cdot g$$

In this equation, something about the units should be mentioned. Usually the model pilot declares the thrust force in grams or kilograms. But technically speaking, the force should be stated in the units of Newton. The value is simply the thrust in kilograms multiplied by the acceleration of gravity g (about 10m/s^2). A quadrocopter which produces 1.2kg of thrust generates, expressed physically correct, 12N of thrust.

In this way, the equation above is also correct. On the left side is for this example 4 x 3N, on the right 1.2kg x 10m/s^2. To compensate for the 12N weight force, each of the four drives must reach 3N thrust force (or in the language of the model constructor 300g or 0.3kg thrust).

With slight modifications of this formula climb flight or descent flight can also be treated. For the climb each drive produces more than 3N thrust, and for descent less.

Climb flight: $4 \cdot F > m \cdot g$, descent flight: $4 \cdot F < m \cdot g$

1.2 The attack angle φ

Next, the pilot or the GPS controller would like to attack the quadro-copter at an angle in the nick or roll direction, so it can drift off to one side and movements in any direction are possible.

How a rotation by the angle φ is actually achieved is looked at later. For the time being, only the forces for the quadrocopter with attack angle φ are considered. Figure 2 shows this from the side. It is the same whether nick or roll is shown, because of the symmetry. Of course, even now the four thrust forces are still active. The middle two shown are from the two propeller shafts of the other axis.

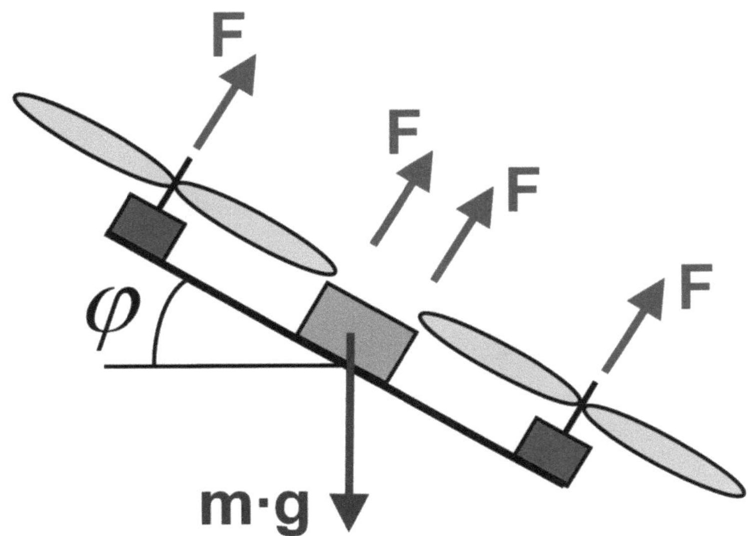

Figure 2: Forces with attack angle φ

The thrust forces are still pointing in the same direction as the motor shaft, and the weight force is still at the center of gravity and points towards the center of the Earth. The forces are no more opposite to each other than it was the case for the pure hover flight.

Forces always act in certain directions; they are so-called vectors. They therefore always have an arrowhead at the end. They can also be decomposed into a sum of other forces. Figure 3 shows how one can understand that. Instead of the force F, two forces can also be drawn. The second starts at the arrowhead of the first one.

F_D is called the drift force, and it acts in the horizontal. The force F_L will be referred to hereinafter as the lift force, and it acts in the vertical. Figure 4 shows an overall arrangement of the decomposed forces. The forces 2 F_L and 2 F_D again belong to the two propellers from the other axis.

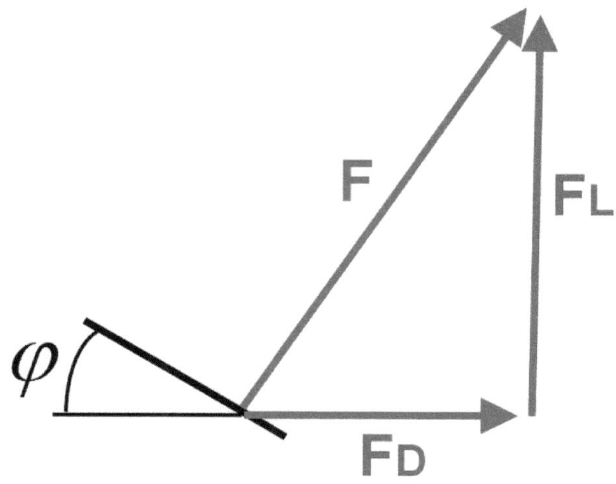

Figure 3: Thrust force F, lift force F_L and drift force F_D

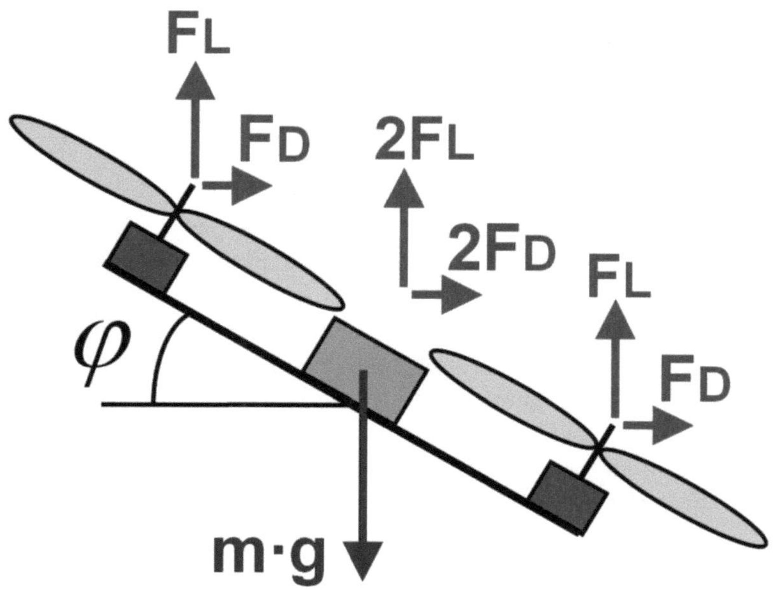

Figure 4: Lift force F_L and drift force F_D

The four lift forces F_L act here again exactly in opposition to the weight force of m · g. To make sure that the quadrocopter holds its height during drifting, the following equation must of course apply again:

$$4 \cdot F_L = m \cdot g$$

The force triangle in Figure 3 is right-angled. It is easy to see that the thrust force F is always the longest side, the hypotenuse. It is always larger than the lift force F_L and larger than the drift force F_D. The position of the throttle stick or the GPS or a height controller always acts on the thrust force F. If the quadrocopter is now attacked with φ, according to the above formula, instead of F, only the smaller lift force F_L acts against the weight force. Without changing the throttle stick, the quadrocopter begins to sink and drifts off to the side (or forward, depending on whether φ acts in the roll or nick direction). The pilot can compensate for this with an increase in throttle position[1].

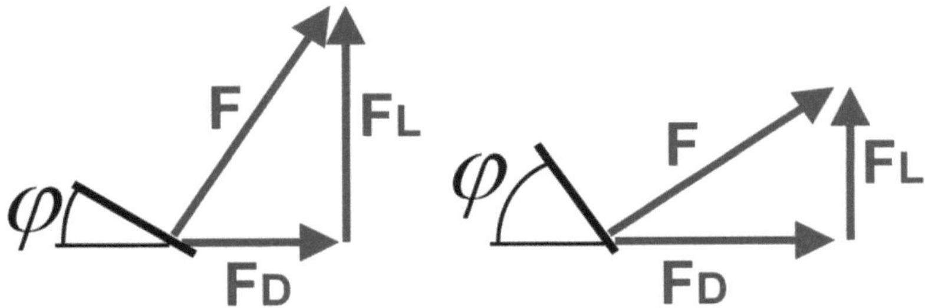

Figure 5: Comparison of two force triangles at different angle φ

Figure 5 shows a comparison as the forces change at a different attack angle φ. It is recognizable that the lift force F_L at a larger attack angle φ and constant thrust F gets smaller in favor of the drift force

---

[1] That's true only if the height controller and the air pressure or other height sensor are turned off. If they are turned on, the force F is set by the controller such that $4 \cdot F_L = m \cdot g$ is always given. For larger φ, the regulator automatically gives more gas to keep the height.

F_D. The quadrocopter drifts faster in the right-hand illustration, and more power must be given to keep it hovering.

Once again, 'actio = reactio' should be investigated. This rule is actually violated in Figure 4. To the four vertical forces F_L, the weight force $m \cdot g$ counteracts. But in the horizontal, it seems that the drift force F_D doesn't counteract anything. If one applies a force F to a body in space, so acts $F = m \cdot a$, Force = mass x acceleration. The body is therefore constantly accelerated by $a = F / m$. Next is indeed $v = a \cdot t$, velocity = acceleration x time. If this would match so all, the drift velocity would increase continuously.

Of course, this does not happen in practice. Indeed, there is something indispensable for quadrocopters; it is also necessary to ensure that the propellers can generate thrust at all – it is air.

He who in a RC-car presses down the throttle stick for a long time notes that the speed at some point no longer continues to increase. This is exactly the same with a quadrocopter. The four drift forces F_D are opposed by an aerodynamic drag force F_A.

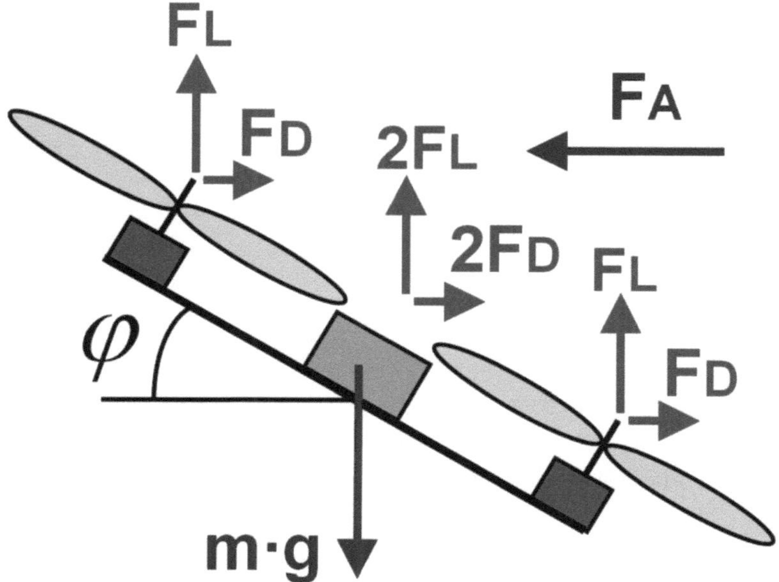

Figure 6: All forces inclusive aerodynamic counterforce

This gets larger with increasing speed, and at some point the same value as 4 · F_D is reached. Then, the maximum velocity is reached. This is illustrated in Figure 6. The aerodynamic counterforce F_A does not attack at a particular point of the quadrocopter. It acts on the entire surface, so it is drawn in the Figure 'free-flying'. In the horizontal, it counteracts the drift forces F_D, so at a certain maximum velocity again the rule 'actio = reactio' applies.

1.3   General balance of forces

The mechanisms that allow the quadrocopter to rotate to the angle φ have not yet been considered. This requires the forces for a general case to be discussed. Without restrictions, it is possible to do this for only two opposing rotors. Figure 7 shows such a general case, in which the force F_1 is not the same size as F_2.

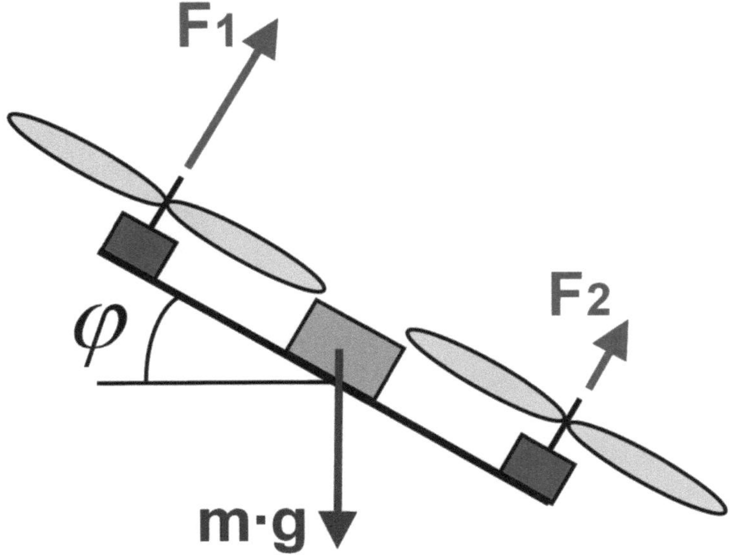

Figure 7: General case: $F_1 \neq F_2$

To imagine what happens in this case, the two forces F_1 and F_2 are each decomposed into two other forces, similarly to the thrust force F being decomposed into the two forces F_L and F_D above. Figure 8 shows this.

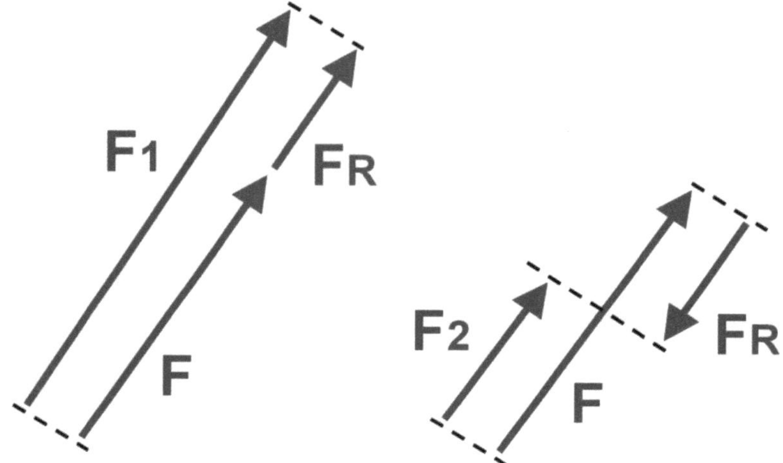

Figure 8: General case, separation of F_1 and F_2 into F and F_R

F_1 and F_2 are decomposed into two forces F and F_R. F is again the previously known thrust force and F_R is a rotational force. Why it is called rotational force is explained below.

It can be seen from Figure 8 that

$$F_1 = F + F_R \qquad \text{and} \qquad F_2 = F - F_R$$

A numerical example will illustrate this. If F = 5N and F_R = 2N, then F_1 = 5N + 2N = 7N and F_2 = 5N − 2N = 3N. Comparing this with the arrow lengths, this also appears plausible. These formulas can also be transformed so that F and F_R can be calculated from given F_1 and F_2:

$$F = \frac{F_1 + F_2}{2} \qquad \text{and} \qquad F_R = \frac{F_1 - F_2}{2}$$

The example numbers from above must also comply with these equations: 5N = (7N + 3N) / 2 and 2N = (7N − 3N) / 2

The effect of the thrust force F was explained above. As shown above, it can just be divided into lift force F_L and drift force F_D. F is therefore responsible for 'lift' and 'drift', but cannot cause any rotation of angle φ.

1.4   A simple physical model

It is thus clear that the force F_R – the rotational force – must be responsible for this rotation. In a simplified view only the effect of the F_R will be further investigated (and it would tacitly be assumed that the quadrocopter still hovers and is possibly drifting due to the not-shown force F).

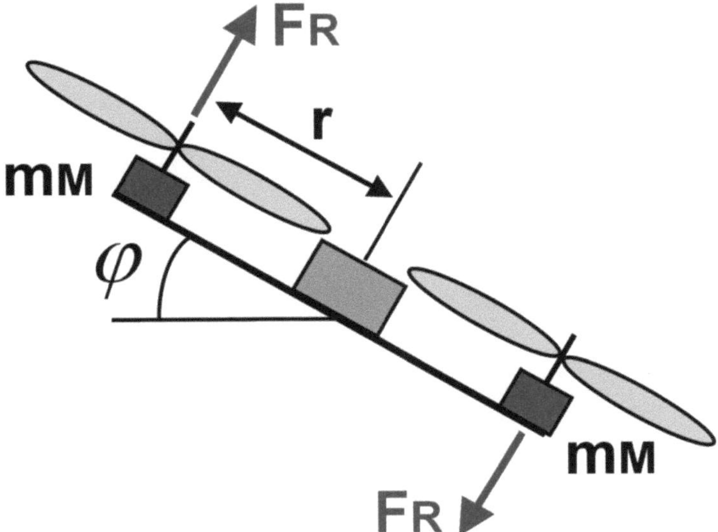

Figure 9: Only F_R is relevant for the rotation around φ

To develop a simple physical model the mass moment of inertia J and the torque M need to be used.

$$a) \qquad M = J \cdot \ddot{\varphi}$$

$$b) \qquad 2 \cdot F_R \cdot r = 2 \cdot m_M \cdot r^2 \cdot \ddot{\varphi}$$

$$c) \qquad \ddot{\varphi} = \frac{F_R}{m_M \cdot r}$$

a.) Means that torque = mass moment of inertia x angular acceleration. This formula is not necessarily familiar to everybody, particularly if the physics class was a long time ago. But it can be derived from F = m · a (force = mass x acceleration). The two formulas are in fact similar; while F = m · a is for movement along an axis, this one refers to rotations around an axis. The so-called angular acceleration $\ddot{\varphi}$ is mathematically the second derivative of the angle.

b.) An illustration of the variables that are shown in Figure 9; $M = 2 \cdot F_R \cdot r$ and $J = 2 \cdot m_M \cdot r^2$ apply. In calculating the mass moment of inertia J, it is assumed that the total mass of the axis consists mainly of the two motor masses m_M.

c.) All of that can be dissolved to $\ddot{\varphi}$ and the result is the following mnemonic:

The rotational force F_R acts to the angular acceleration $\ddot{\varphi}$.

The quadrocopters balance the angle automatically. The pilot gives only a desired angle φ by the nick or roll stick of the remote control transmitter or in todays fully autonomous systems, the GPS control gives the desired angle. This angle is achieved afterwards through internal control. The regulators of this control are discussed in the next chapter, 'Axis Control'.

The integration of the angular acceleration is the angular velocity and the integration of the angular velocity is the angle itself. In order to calculate the angle φ, the angular acceleration $\ddot{\varphi}$ must therefore be integrated twice. At the very end of the chapter this is described in a graphical representation. The two integrators can also be presented with the equivalent description using Laplace by 1/s.

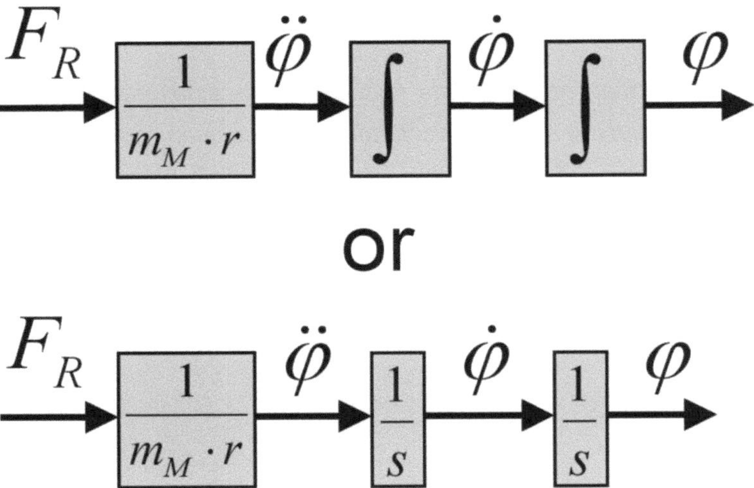

Figure 10: A simple physical model of one axis

1.5 Findings in brief

Here are the key findings of the chapter shown again in brief.

For the pure floating $4 \cdot F = m \cdot g$ applies. The addition of the four thrust forces F acts against the weight force and keeps the model in balance.

If the quadrocopter now is attached at φ, according to the above formula, instead of F, only the smaller lift force F_L acts against the weight force: $4 \cdot F_L = m \cdot g$, but it drifts off to the side. The pilot can compensate for this with an increase in throttle position.

To make a rotation through the angle φ possible, the forces F_1 and F_2, generated by two opposing propellers, should not be the same. This results into a rotational force F_R, which is solely responsible for the rotation.

The rotational force F_R acts to the angular acceleration $\ddot{\varphi}$. The integration of the angular acceleration is the angular velocity and the integration of the angular velocity is the angle itself. In order to calculate the angle φ, the angular acceleration $\ddot{\varphi}$ must therefore be integrated twice.

2   Axis Control

The setting of the controller leads to great discussions in Internet chat forums. The discussions concern the question of how the controller parameters or 'settings' must be chosen so that a certain desired flight performance results. This chapter aims to create a common basis for all systems. Therefore, the large number of setting parameters from the various software vendors need to be discussed at the beginning.

On the following pages, only the control of the nick and roll axis is treated. The parameters which are responsible for that – K_P and K_D – are always the most important of the whole system, as they are directly responsible for the behavior, such as agility for aerobatics pilots and good-natured for the camera and hovering pilots.

Regulators which contain K_P and K_D are called proportional differential (PD) controllers. K_P is also called the 'P-part', and K_D the 'D-part'.

2.1   Control of nick and roll axis

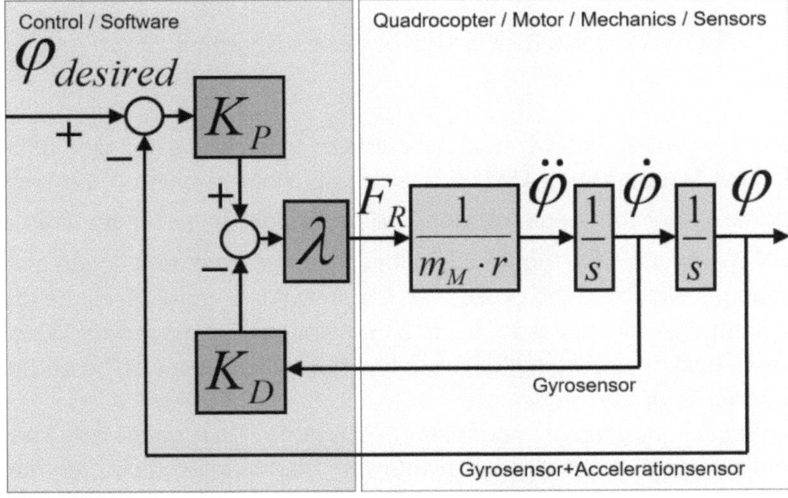

Figure 11: Control of nick and roll axis, block diagram

Figure 1 shows a block diagram of the controlled nick or roll axis. The right side represents the hardware of the quadrocopter. The core forms the physical model of the axis, developed in the last chapter. The left side shows how the parameters of the PD controller are linked to it. It has to be mentioned that the left side is software, and the parameters K_P and K_D also have to be set in the software. The angular velocity $\dot{\varphi}$ is measured by the gyro sensor, and the angle φ with the combination of gyro sensor and acceleration sensor.

The associated transfer function of the system is shown in Figure 12.

$$\frac{\varphi}{\varphi_{desired}} = \frac{K_P \cdot \dfrac{\lambda}{m_M \cdot r}}{s^2 + s \cdot K_D \cdot \dfrac{\lambda}{m_M \cdot r} + K_P \cdot \dfrac{\lambda}{m_M \cdot r}}$$

Figure 12: Transfer function of block diagram of Figure 11

The combination of gyro and acceleration sensor is also called IMU, Inertial Measurement Unit. If it is not possible to pick up the gyro signal $\dot{\varphi}$ for the feedback loop, the controller has to be modified a bit. This is shown in Figure 13. In this case, the transfer function also changes slightly in the counter and is shown in Figure 14. Since the D component in this case has to be multiplied by the operator "s" for the derivation according to time, a low-pass filter must also be added in practice at the implementation.

In most quadcopter controllers, complete PID controllers are implemented. In the case of plants with double integrators, as they occur in quadcopters from the force to the angle, integrators or integral components of the controller have a destabilizing effect. But they can be useful with constant interference. However, when setting parameters, you should ensure that the I components are kept very small in order to reach a robust stability of the system.

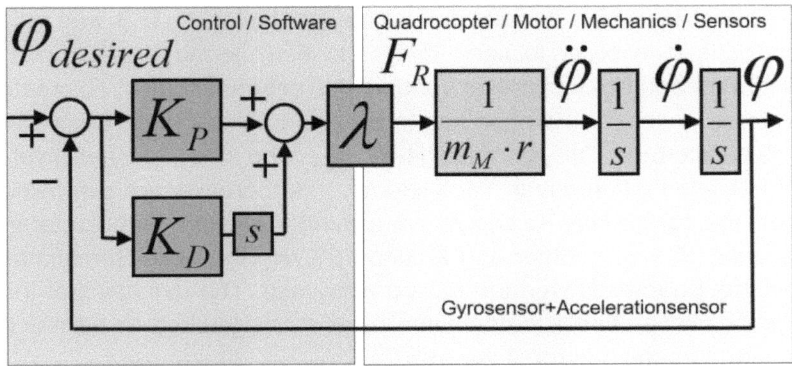

Figure 13: Control of nick and roll axis, block diagram

$$\frac{\varphi}{\varphi_{desired}} = \frac{s \cdot K_D \cdot \dfrac{\lambda}{m_M \cdot r} + K_P \cdot \dfrac{\lambda}{m_M \cdot r}}{s^2 + s \cdot K_D \cdot \dfrac{\lambda}{m_M \cdot r} + K_P \cdot \dfrac{\lambda}{m_M \cdot r}}$$

Figure 14: Transfer function of block diagram of Figure 13

The model pilot or the GPS controller commands the desired angle with the nick and roll stick of the remote control. The desired angle $\varphi_{desired}$ is supposed to be compared with the true angle φ. Therefore the true angle is subtracted from the desired angle and the result is multiplied by a factor K_P (for K proportional). K_P will have the task of influencing the rotational force F_R so that the desired angle and the true angle are the same after a while.

The signal of the gyro is the time derivative of the angle. This signal is amplified by a factor K_D (for K differential). Afterwards, it is also acting onto the force F_R. One could prove that the angle could not be stabilized without the additional gyro signal and the factor K_D or at least the calculation of the K_D path after Figure 13. This will be overlooked for now.

There still remains the question of the last factor, λ. This was only introduced in order to keep things general, because electronics, software and flight controller vendors use different scales. To create a force, a speed reference has to be sent to a brushless controller via an interface. This is done digitally, so on the 'bits and bytes level'. It is scaled differently in the systems. It is therefore not surprising that the parameters K_P and K_D are around 1 with one manufacturer, around 10 with another and around 100 with the next. Sometimes different names for K_P and K_D are also used. The dynamics of the motor and propeller is neglected here, it is assumed to be much faster than the control of the axis.

2.2   Effect of K_P and K_D

Thus, it is now possible to describe exactly how different values of K_P and K_D effect the behavior of the quadrocopter. This will not be carried out 'by hand', but with a simulation program. There, the block diagram of Figure 11 or the Transfer Function of Figure 12 are simply entered and the curves of the angle can then be easily displayed.

The following simulations are only examples of how to find meaningful values, the values cannot be transferred 1:1 to any existing flight controller. If you simulate the two block diagrams in the Figures 11 and 13, the behavior is different, so it is important that you map the parameters and block diagrams of the systems correctly when you configure your own flight controller or quadcopter. The simulations are made for a standard quadrocopter. The data are about: motors, m_M = 0.030kg to 0.060kg, spacing of axes, r = 0.01m to 0.03m. It is always assumed that, at the beginning, the system is hovering and the pilot or the GPS control suddenly commands a desired nick angle of 10° via the remote control.

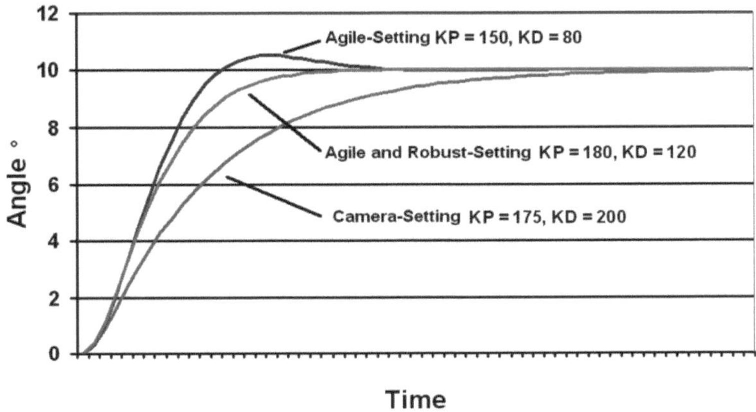

Figure 15: Angle curve φ for different values of K_P and K_D

Figure 15 shows some angle curves for different settings. These are named in the following according to their behavior. The 'Camera Setting' displays a very good-natured behavior. The angle is slowly approaching 10°. If the pilot makes a steering error, this only has an effect on the quadrocopter after a time delay. Therefore counter-steering is still easily undertaken.

A comparison of the 'Camera Setting' with 'Agile Setting' at 8°, i.e. when almost the full scale of 10° is achieved, shows something interesting. Using 'Agile Setting', the controller requires only about half as much time to reach this angle. This also means that steering mistakes become apparent here much more quickly.

This 'Agile Setting' has yet another feature: a so-called overshoot. At one point, the angle is slightly greater than 10°, before it then falls back to that value. This is normal. Systems in which the control parameters are set too 'fast' very often show an overshoot.

A model pilot can also see these characteristics of the settings in his aircraft. For that purpose he pushes the stick of the remote control back and forth while the quadrocopter is hovering.

Using 'Camera Setting' it can be observed that the quadrocopter 'is not hanging on the stick', but that the achieved angle always lags something behind to the desired angle. This is shown in the graph

21

in Figure 15. The angle of 10° desired by the pilot is achieved far more slowly than in other settings.

This is of course nothing for a speed flying pilot, but ideal for camera or hover flights.

Using 'Agile Setting', the quadrocopter 'hangs more on the stick'. An ideal 'hanging on the stick' would be that all movements of the remote control's stick can be observed immediately at the aircraft. This is not possible because the drives can't change the speed fast enough, the thrust forces are limited, the transmission of the remote control is too slow, etc. The speed and aerobatic pilot is nonetheless happy, because everything he does on the remote control quickly arrives at the model. However, the camera pilot doesn't like that setting at all, because steering errors of course also arrive quickly.

In Figure 15, a third setting is illustrated, which is called 'Agile and Robust Setting'. It is a compromise between the other two. It is on the one hand almost as agile as the 'Agile Setting' but has, on the other hand, no overshoot. Of course, the author has also tested this setting on his quadrocopter, and found it as stable as piece of wood in the air that can still act as agile as a Ferrari.

Squirrelly and fidgety

But it is not only the steering mistakes which become noticeable more quickly for the one setting and slower for the other. Exactly the same applies to interference from wind or noise signals from the sensors. Therefore, a 'Camera Setting' quadrocopter is smoother in the air, also in a hover flight, with no additional commands.

This fact also helps the camera pilot. To keep the camera still when photographing is ultimately the most important measure against blurred images. In contrast, a system in the 'Agile Setting' behaves much more squirrelly and fidgety. It immediately tries to steer a disturbing influence, and furthermore it overshoots.

Changing K_D while keeping K_P the same

Anyone who has ever tried to adjust the K_P and K_D with a PD controller for a desired behavior has realized that this is not such a simple task. There are just two parameters which are independent, and therefore there are almost endless possibilities. To investigate the influences systematically, the following procedure is pursued in Figures 16 and 17.

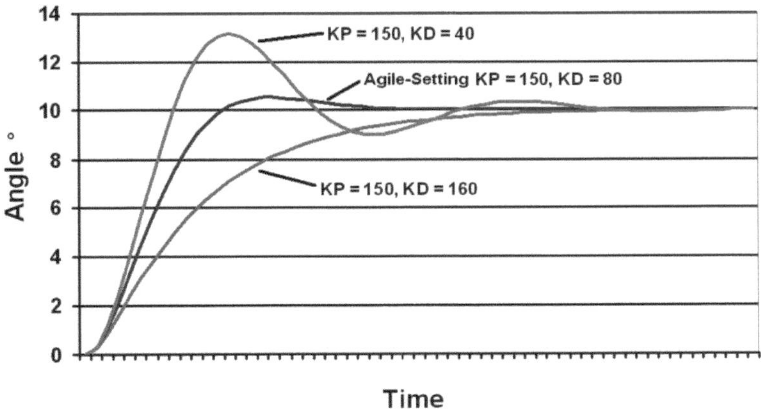

Figure 16: Angle curve φ by changing K_D

It is always started with the 'Agile Setting'. Then, one of the two parameters remains the same and the other is changed.
Figure 16 shows a change of K_D, while K_P always remains at 150. One can clearly see that the overshoot is very large with too small a K_D, while the system is getting even more agile. With those settings, the quadrocopter would actually be almost un-flyable. An increase in the K_D, however, makes the system good-natured.
If K_D remains at 80 and K_P is changed as shown in Figure 17, the system behaves with a K_P of 200 a little more agile and for a smaller K_P good-natured again.

Generally the following observations can be made:

For smaller K_P and larger K_D, the system is good-natured, slow and less susceptible to interference.
For larger K_P and smaller K_D, the system is more agile and responds quickly to disturbances.
If K_D is too small, the system is less damped and therefore tends to overshoot.

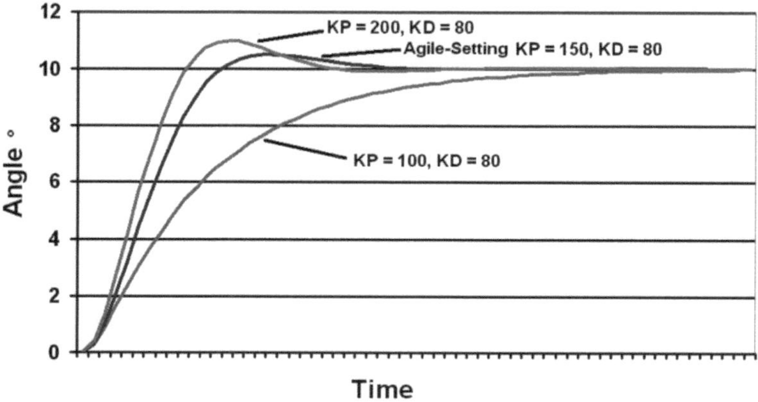

Figure 17: Angle curve φ by changing K_P

Best practice by setting the controller

If the quadrocopter is built with the motors, propellers and spacing of axes as recommended by the manufacturer, a suggested good-natured beginner setting would allow a stable levitation. The further offered settings as example 'Sport', 'Agile', 'Advanced', 'Expert', 'Camera', 'Beginner', and the rest of them – often provide sufficient elbowroom for most users and they can be easily adopted. If a quadrocopter doesn't hover with the suggested components and settings, it is often not the K_P and K_D parameters but other aspects that are responsible.

Things get more interesting when other motors, propellers and spacing of axes are used.

Figures 16 and 17 show very nicely how the overall behavior is changed with varying K_P and K_D. In practice, however, the model pilot doesn't have such graphs for 'his' aircraft on hand. He sets the parameters with the PC interface, starts the motors and can only find out visually whether it behaves as desired. So he must 'aviate' the parameters.

Therefore he starts with a good-natured 'beginner setting'. If the quadrocopter hovers, the optimization can be started, depending on the desired behavior.

Agile behavior: Decrease K_D or enlarge K_P, until an overshoot is visible. You see that at first the angle deflects further than at the end. If the system already overshoots strongly while running the startup parameters, then enlarge K_D a little and decrease K_P.

Camera behavior: Enlarge K_D or decrease K_P until the quadrocopter is visibly lagging behind, i.e. no longer 'hanging on the stick'.

Another approach may be: starting from a 'Beginner Setting', enlarge the value of K_P and decrease K_D by the same factor to effect more agility, or decrease the value of K_P and increase the K_D by the same factor to cause more good-natured behavior.

2.3 Transfer function

To learn something more about the influence of parameters, and to provide even more setting tips, the transfer functions of Figures 12 and 14 can also be examined more closely.

The 's', which can be found in the denominator, may allow some dynamic studies to be made using this transfer function. The factor 'λ' was discussed in Section 2.1. 'm_M' and 'r' are provided in Figure 9. For the considerations at this point, special attention is to be directed to only a single key issue: The transient behavior of the angle and thus the behavior of the quadrocopter remains exactly the same if all terms of the equation also remain the same[2].

If you now want to change things structurally and find new values for K_P and K_D so that the behavior remains the same as before, you must only ensure that

$$K_P \cdot \frac{\lambda}{m_M \cdot r} \quad \text{and} \quad K_D \cdot \frac{\lambda}{m_M \cdot r}$$

remain the same.

1st example: The spacing of the axes will be increased from 45 cm to 60 cm. Otherwise everything remains the same. r is in the denominator in both terms and therefore changes them to 1 / (60/45), so both K_P and K_D must be greater by 60/45. The new values are K_P = 150 x 60/45 = 200 and K_D = 80 x 60/45 = 107.

2nd example: Smaller motors of 30g are installed. Thus, this changes both terms by 1/(30/60). K_P and K_D would have to be reduced by half. But in addition, the two drives generate only about half the thrust[3]. This means that at a certain speed only about half as much thrust is available. Thus, λ is also approximately reduced by half. K_P and K_D would therefore have to be doubled to compensate again. The new values are K_P = 0.5 x 2 x 150 = 150 and K_D = 0.5 x 2 x 80 = 80. The values can therefore be left unchanged.

Example 2 is also the reason why so many pilots realize a similar axis distance of maybe 35cm to 45cm and fly with the same controller settings with comparable properties, even though they are installing different motors and propellers with different power.

---

[2] Terms for the present example: $K_P \cdot \dfrac{\lambda}{m_M \cdot r}$ and $K_D \cdot \dfrac{\lambda}{m_M \cdot r}$

[3] This is only rudimentarily true, since the relation between speed and thrust is not linear.

Therefore, in various Internet forums, parameter settings are compared and discussed, even though many different motors and propellers are used. Because a heavier drive also provides more thrust, the influence on the control parameters K_P and K_D is minimal.

2.4 Findings in brief

As in the last chapter, here the main findings are presented in abbreviated form.

The parameters K_P and K_D are always the most important of the whole system. They are used for the balancing of the nick and roll axis.
The manufacturers of quadrocopters often use different names for those parameters.

With smaller K_P and larger K_D, the system is good-natured, slow and less susceptible to interference. With larger K_P and smaller K_D, the system is more agile and responds quickly to disturbances.

If K_D is too small, the system is less damped and therefore tends to overshoot.

An easy approach for parameter adjustment may be: Starting from a 'Beginner Setting', to enlarge the value of K_P and decrease K_D by the same factor to effect more agility, or decrease the value of K_P and enlarge the K_D to the same factor to cause more good-natured behavior.

Because a heavier drive also provides more thrust, the influence on the control parameters K_P and K_D is minimal.

3 GPS Control

3.1 Basic principle of route control

Flying to a point in space is a three-dimensional problem. However, for the sake of simplicity, it is only described in one dimension. If a quadrocopter wants to approach a point in space that is in a horizontal line away from it, it is necessary that it moves transversely with an angle of attack φ. Figure 18 shows the forces of a quadrocopter with and without an angle of attack, as already discussed in chapter 1.

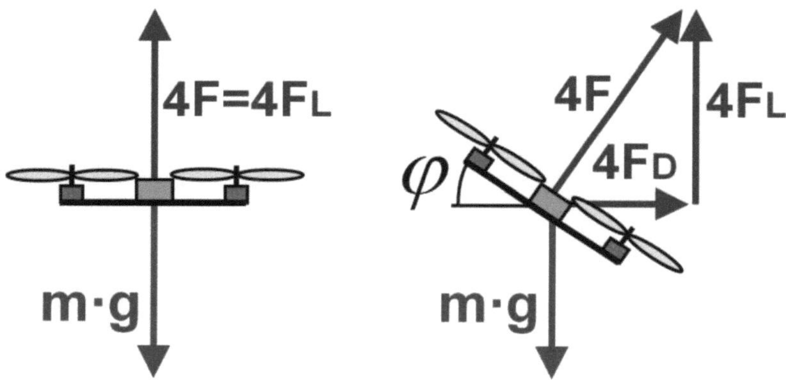

Figure 18: left: stable state of hover flight, right: drift to the right with angle of attack

The force F is again the thrust that is caused by a propeller. It can only act perpendicular to the quadrocopter position, since the propellers rotate with the quadrocopter. A total of $4 \cdot F$ is effective, since there are four propellers. The picture on the left shows the stable state of hover flight. The forces are in equilibrium here, so $4 \cdot F = m \cdot g$ applies. The total thrust here also acts as the lift force $4 \cdot F_L$. It can also be seen from the illustration that otherwise no further forces act. As mentioned above, no horizontal movement is possible.

On the right in the picture the quadrocopter was set up at the angle φ. The thrust can now be divided into a horizontal and a vertical component. One can also easily imagine that instead of the thrust force F, a drift force F_D and a lift force F_T are now acting. The drift force now causes the system to move to the right.

This illustration also shows another property of the quadrocopter attached to an angle φ. To do this, consider the right-angled triangle that is spanned by F, F_D and F_L, as already discussed in chapter 1. F is always the hypotenuse, i.e. the longest of the sides. If a quadrocopter is attached to an angle, then only the smaller lift force F_L counteracts the weight m · g. A drifting quadrocopter always loses altitude if the thrust is not increased accordingly.

GPS control is therefore always accompanied by altitude control via an altitude sensor such as an air pressure sensor. This ensures that the thrust F is increased accordingly in the event of larger drift in order to maintain the flight altitude.

3.2 Dynamics of drifting

How this behaves in the dynamics is shown in Figure 19. It shows the course of the angle of attack. At time t_1 the quadrocopter is in a stable hovering position. At a later point in time t_2, it is likewise again in a stable hovering position, which, however, is locally further to the right.

Figure 19: Dynamics of drifting

To do this, the angle of attack is first increased. This is achieved with a control that measures the angle with the combination of a gyro and acceleration sensor or with the IMU, as described in chapter 2. At a certain point in time, the maximum possible angle is reached and the quadrocopter drifts with the maximum drift force F_D. In order to achieve a stable floating position again at B, the angle of attack is then reduced again. But that alone is not enough. In the middle of the picture, the angle of attack and thus also the drift force are zero again. However, since the drift force always acted in the same direction before, the speed is then not zero and the quadrocopter continues to drift.

A little physics should help here. It is true that "force = mass x acceleration" or "acceleration = force / mass" applies. Furthermore, the following also applies: "speed = acceleration x time" or also "speed = force / mass x time".

Strictly speaking, this only applies to constant values, which is actually not the case here, since the force changes with the angle of attack. However, the fact is that even when the force has changed, it has always been acting in the same direction. Therefore a countermovement has to be carried out again. The angle of attack must therefore deflect again to the other side. This means that a force can act in the other direction and brake the quadrocopter again, so the speed will be zero again in the end. For this consideration, the air resistance was neglected. In practice, this also decelerates the drifting of the quadrocopter and the countermovement is therefore always smaller or not as long as shown in the picture. It goes without saying that the dynamics of drifting is quite a complicated matter. A control loop is therefore needed that does this task.

This process can only be examined more closely if the whole thing is viewed from the side of physics. In the following, a physical model should therefore be created again in the context with the GPS control of the quadcopter.

3.3   Physical model

Figure 20 shows again the triangle of forces of the quadrocopter attached with the angle φ. It should now be examined more in detail.

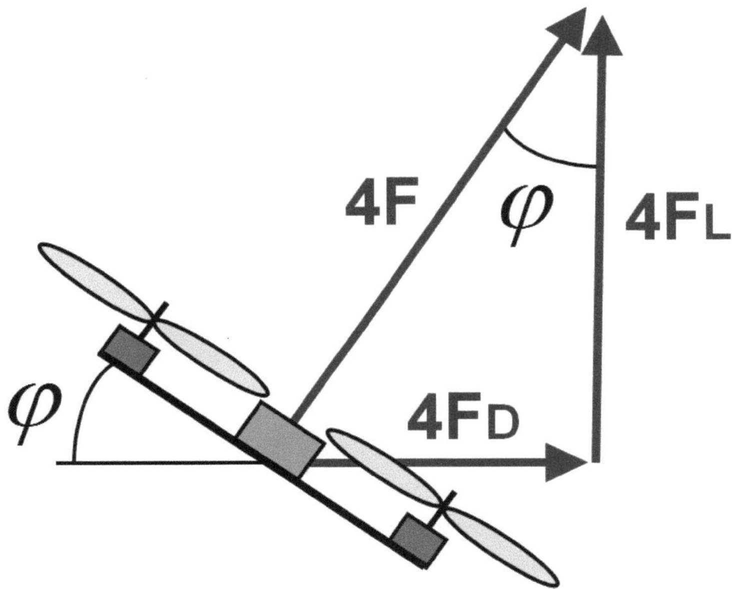

Figure 20: Triangle of forces

The angle of attack φ can be found again between F and F$_L$. The trigonometry then describes the length ratios in a right-angled triangle

$$\frac{4 \cdot F_D}{4 \cdot F} = \sin \varphi \qquad \text{or} \qquad F_D = F \cdot \sin \varphi$$

The factor 4 is reduced here. The sine function behaves almost linearly for small angles and it then also applies approximately

$$\sin \varphi \approx \varphi \qquad \text{or used in the formula} \qquad F_D \approx F \cdot \varphi$$

You can easily check with the pocket calculator that the sin φ is approximately equal to φ for small angles if you set it to RAD.

Together with the formula "force = mass x acceleration" used above, a simple physical model can now be derived. The force acting in the horizontal axis is F_D and the mass is that of the quadrocopter, m. The acceleration is then a. So the following applies:

$$F \cdot \varphi \approx F_D = m \cdot a \qquad \text{or resolved according to a:} \qquad a \approx \varphi \cdot \frac{F}{m}$$

The acceleration a is roughly proportional to the angle of attack φ. It is known from physics that the acceleration a, integrated with time, results in the velocity v. If the speed v is once more integrated with the time, the covered distance x results. Here, too, the air resistance is neglected and it is assumed that the angle of attack φ is small.

3.4   Angle φ, acceleration a, velocity v and distance x

With this little excursion into physics, Figure 19 can now also be better understood. Therefore Figure 21 shows these sketches again. This time, however, they are supplemented with the angle of attack φ.

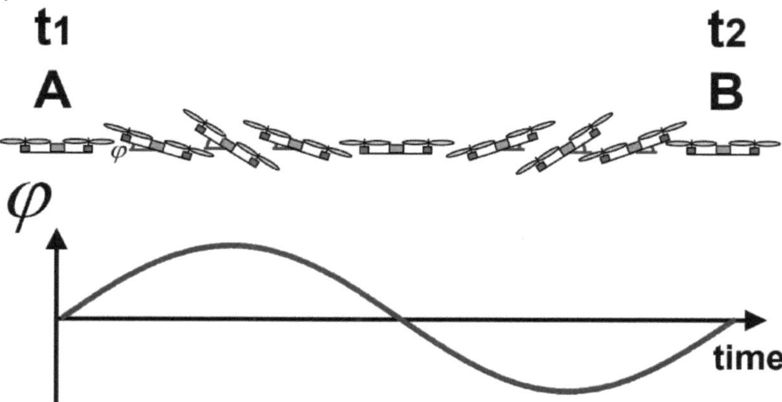

Figure 21: Change in the angle of attack over time

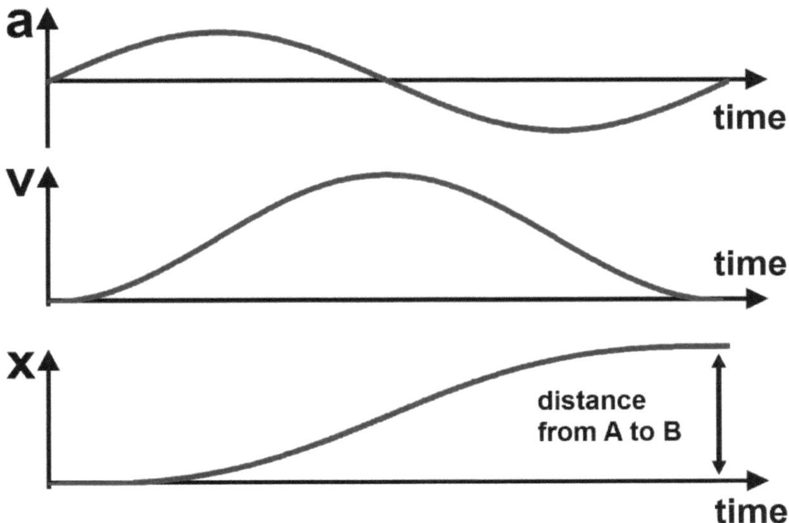

Figure 22: acceleration a, velocity v and distance x

As can be seen, the angle of attack is first getting larger, then smaller again, in order to become zero in the center of the image. When moving in the opposite direction, the angle is pointing in the opposite direction. This is shown with a negative angle.

If this course of the angle of attack acts on the quadrocopter, also the courses of a, v and x result. The acceleration a is calculated as described above as $a = \varphi \cdot F / m$. F has to be slightly larger for drifting in order to maintain the height, but this effect is negligible. Thus, a is roughly proportional to φ. The speed v and the distance x are then each the signals integrated with respect to time. As described, the result is a path from A to B.

3.5 Control

In chapter 2 it was described that a PD controller with the parameters K_P and K_D is required so that the angle of attack φ follows a specified angle. If the quadrocopter is operated without GPS, the pilot simply specifies this angle with the stick position for nick or in the other axis for roll. If the GPS is switched on, a control takes over the task of determining the angle of attack φ for the two axes.

The angle control from chapter 2 is completely embedded in the new system and is a central component of it. Figure 23 shows this. Here, for horizontal flight, the force F is about m·g for the assumed small angle φ.

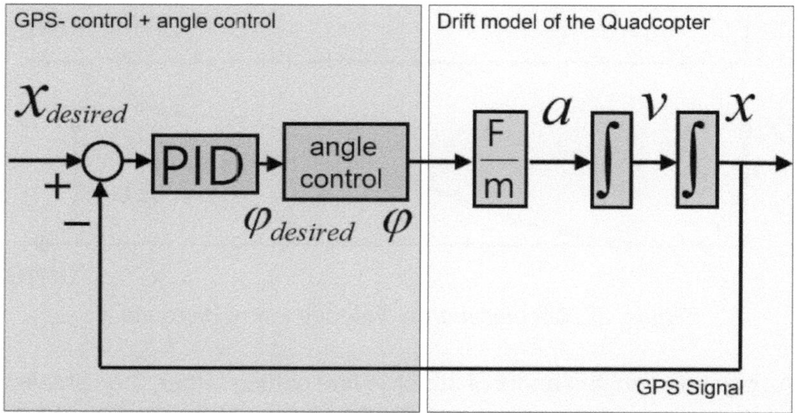

Figure 23: GPS control combined with angle control of chapter 2

Now, however, there is no longer a desired angle of attack specified, but a value $X_{desired}$. $X_{desired}$ stands for any point in space. Of course, the same also exists in the Y direction. For these considerations, it is assumed that the nick and roll axes each act in the same direction as the X and Y directions provided by the GPS. In order to ensure that the direction is always maintained, the compass sensor and the yaw angle control must be switched on. This leads to the following mnemonics:

A GPS control only works reliably if both the control of the yaw angle via the compass sensor and the altitude control via the air pressure sensor are switched on.

The regulation of the yaw angle is important because the GPS controller must specify the angle of the correct axis (nick and roll) for each of the X and Y axes. The height control compensates for the thrust when drifting so that the quadrocopter does not lose any height.

In practice it is also conceivable that the nick and roll axes are rotated to the GPS X and Y. Even then, the GPS controller can specify the correct angles via a coordinate transformation. It is important, however, that the angle at which the nick and roll axes are rotated relative to the X and Y axes is always known in order to be able to carry out this coordinate transformation correctly.

3.6   Influence of wind

The influence of the wind that is almost always present in practice must also be taken into account in the drift model of the quadrocopter. Using some theory, it can be shown that a good GPS control, which also controls the disturbance of the wind, can be realized with a PID controller (Proportional - Integral - Differential). The P and D component are again responsible for stability and the I component should be kept small. This can be set in the configuration software via the PC interface, or directly programmed into the flight controller. The parameters can be called GPS-K_P, GPS-K_D and GPS-K_I or similar, for example. The manufacturers of flight controllers for quadrocopters already make suggestions for GPS-K_P, GPS-K_D and GPS-K_I in their basic settings, as well as for the values for angle control, K_P and K_D. With this, the GPS-controlled quadrocopter should already reach the desired position with some degree of accuracy.

4   Literature

(1) Integrierte Navigationssysteme. Sensordatenfusion, GPS und Inertiale Navigation, Jan Wendel, Verlag Oldenbourg 2007, ISBN: 978-3-486-58160-7

(2) Global Positioning Systems, Inertial Navigation, and Integration, Mohinder S. Grewal, Lawrence R. Weill, Angus P. Andrews, Verlag: John Wiley & Sons, Inc. 2007, ISBN: 978-0-470-04190-1

(3) Das Depron Buch, Hinrik Schulte, Verlag für Technik und Handwerk GmbH 2008, ISBN 978-3-88180-741-8

(4) Moderne Fernsteuerungen für RC-Flugmodelle, Manfred-Dieter Kotting, Verlag für Technik und Handwerk GmbH, ISBN: 978-3-88180-780-7

(5) Büchi, Roland. "Brushless-Motoren und–Regler." *Baden-Baden: Verlag für Technik und Handwerk neue Medien GmbH* (2013).

(6) Selbstbau von Brushless-Aussenläufer-Motoren, Heinrich Hilgers, Neckar-Verlag 2006, ISBN 978-3-78830-683-0

(7) Regelungstechnik, Einführung in die Methoden und ihre Anwendung, Otto Föllinger, Hüthig Verlag 2008, ISBN: 978-3-7785-2970-6

(8) Büchi, Roland. *Radio control with 2.4 GHz*. BoD–Books on Demand, 2014., ISBN 978-3732293407